YOUR KNOWLEDGE HAS VALUE

- We will publish your bachelor's and master's thesis, essays and papers

- Your own eBook and book - sold worldwide in all relevant shops

- Earn money with each sale

Upload your text at www.GRIN.com and publish for free

Bradley Tice

A Compression Program for Chemical, Biological, and Nanotechnologies

GRIN Verlag

Bibliografische Information der Deutschen Nationalbibliothek:

Die Deutsche Bibliothek verzeichnet diese Publikation in der Deutschen National-
bibliografie; detaillierte bibliografische Daten sind im Internet über http://dnb.d-
nb.de/ abrufbar.

Imprint:

Copyright © 2008 GRIN Verlag GmbH
Druck und Bindung: Books on Demand GmbH, Norderstedt Germany
ISBN: 978-3-656-64522-1

This book at GRIN:

http://www.grin.com/en/e-book/198602/a-compression-program-for-chemical-bio-
logical-and-nanotechnologies

GRIN - Your knowledge has value

Der GRIN Verlag publiziert seit 1998 wissenschaftliche Arbeiten von Studenten, Hochschullehrern und anderen Akademikern als eBook und gedrucktes Buch. Die Verlagswebsite www.grin.com ist die ideale Plattform zur Veröffentlichung von Hausarbeiten, Abschlussarbeiten, wissenschaftlichen Aufsätzen, Dissertationen und Fachbüchern.

Visit us on the internet:

http://www.grin.com/

http://www.facebook.com/grincom

http://www.twitter.com/grin_com

A Compression Program for Chemical, Biological, and Nanotechnologies

By Bradley S. Tice
Advanced Human Design, P.O. Box 3868, Turlock, California 95381 U.S.A.

Abstract

The paper will introduce a compression algorithm that will use based number systems beyond the fundamental standard of the traditional binary, or radix 2, based system in use today. A greater level of compression is noted in these radix based number systems when compared to the radix 2 base as applied to a sequential strings of various information. The application of this compression algorithm to both random and non-random sequences for compression will be reviewed in this paper. The natural sciences and engineering applications will be areas covered in this paper.

Keywords: Compression Algorithm, Chemistry, Biology, and Nanotechnology

I. Introduction

A binary, or radix 2 based, system is defined as two separate characters, or symbols, that have no semantic meaning apart from not representing the other character. This is the same notion Shannon gave to the binary based system upon it's publication in 1948 [1]. This paper will present research that shows how various radix based number systems have a compression value greater than the traditional radix 2 based system as in use today [2]. The compression algorithm will be used to compress various random and non-random sequences. The work has applications in theoretical and applied natural sciences and engineering.

2. Randomness

The earliest definition for randomness in a string of 1's and 0's was defined by von Mises, but it was Martin-Lof's paper of 1966 that gave a measure to randomness by the *patternlessness* of a sequence of 1's and 0's in a string that could be used to define a random binary sequence in a string [3 and 4]. A non-random string will be able to compress, were as a random string of characters will not be able to compress. This is the classical measure for Kolmogorov complexity, also known as Algorithmic Information Theory, of the randomness of a sequence found in a binary string.

3. Compression Program

The compression program to be used has been termed the *Modified Symbolic Space Multiplier Program* as it simply notes the first character in a line of characters in a binary sequence of a string and subgroups them into common or like groups of similar characters, all 1's grouped with 1's and all 0's grouped with 0's, in that string and is assigned a single character notation that represents the number found in that sub-group, so that it can be reduced, compressed, and decompressed, expanded, back to it's original length and form [5]. An underlined 1 or 0 is usually used to note the notation symbol for the placement and character type in previous applications of this program. The underlined initial character to be compressed will be used for this paper.

4. Application of Theory

The compression algorithm will be used for the following radix based number systems: Radix 6, Radix 8, Radix 10, radix 12 and radix 16. These are traditional radix base numbers from the field of computer science and have strong applications to other fields of science and engineering due to the parsimonious nature of these low digit radix base

1

number systems [6].   The compression algorithm in this paper can be both a 'universal' compression engine in that all members of a sequence, either random or non-random, can be compressed or a 'specific' compression engine that compresses only specific types of sub-groups within a random or non-random string of a sequence.

The compression algorithm will be defined by the following properties:

1.) Starting at the far left of the string, the beginning, and moving to the right, towards the end of the string.

2.) Each sub-group of common characters, including singular characters, will be grouped into common sub-groups and marked accordingly.

3.) The notation for marking each sub-group will be underling the initial character of that common sub-group.   The remaining common characters in that marked sub-group will be removed.   This results in a compressed sequential string.

4.) De-compression of the compressed string is the reverse process with complete position and character count to the original pre-compressed sequential string.

5.) This will be the same processes for both random and non-random sequential strings.

.

5. Chemistry

Chemistry is the science of the structure, the properties and the composition of matter and it's changes [7].

5.1 Polymer

A polymer is macromolecule, large molecule, made up of repeating structural segments usually connected by covalent chemical bonds [8].

5.2 Copolymer

A copolymer, also known as a heteropolymer, is a polymer derived from two or more monomers [9].

Types of Copolymers;

1.) Alternating Copolymers: Regular alternating A and B units.

2.) Periodic Copolymers: A and B units arranged in a repeating sequence.

3.) Statistical Copolymers: Random sequences.

4.) Block Copolymers: Made up of two or more homopolymer subunits joined by covalent bonds.

5.) Stereoblock Copolymer: A structure formed from a monomer.

An example of the use of a compression algorithm on copolymers is as follows:

1.) Alternating Copolymers: Alternating copolymers using a radix 2 base number system.

Unit A = 0

Unit B = 1

Example # 1:

01010101010101

Compression of Example # 1

<u>Key Code</u>

0 = 7 characters

1 = 7 characters

Example # 1 Compressed

<u>01</u>

The compressed state of Example #1 is a 2 character length from the original non-compression state total of 17 characters in length.

Periodic Copolymers: Periodic copolymers using a radix 16 base number system.

Unit A = abcdefghijklmnop

Unit B = 123456789@#$%^&*

Example #2

abcdefghijklmnop123456789@#$%^&*123456789@#$%^&*abcdefghijklmnop123456789@#$%^&*

Compression of Example #2

<u>Key Code</u>

abcdefghijklmnop = 16 characters

123456789@#$%^&* = 16 characters

Example #2 Compressed

<u>a11a1</u>

The compressed state of Example #2 is 5 characters from the original non-compression state total of a 80 character length.

Statistical Polymers: Random copolymer using a radix 8 base number system.

Unit A = 12345678

Unit B = abcdefgh

Example #3

12345678abcdefghabcdefgh1234567812345678abcdefgh123456781234567812345678

Key Code

12345678 = 8 characters

abcdefgh = 8 characters

Compression of Example #3

1a11a111

The compressed state of   Example #3 is 8 from the original non-compression state total of a 64 character length.

Block Copolymers: Block copolymer using a radix 12 base number system.

Unit A = abcdefghijkl

Unit B = 123456789@#$

Example #4

123456789@#$123456789@#$123456789@#$abcdefghijklabcdefghijklabcdefghijkl

Key Code

abcdefghijk = 12 characters

123456789@#$ = 12 characters

Compression of Example #4

111aaa

The compressed state of Example #4 is 6 characters from the original non-compression state of 58 character length.

Stereoblock Copolymer: Stereoblock copolymer using a radix 10 base number system.

Unit A = abcdefghij

Unit B = 123456789@

Note: The symbol [I]   represents a special structure defining each block.

Example #5

abcdefghijabcdefghijabcdefghijabcdefghijabcdefghijabcedfghij
                                      I   I
123456789@123456789@   123456789@123456789@

Key Code

abcedfghij = 10 characters

123456789@ = 10 characters

Compression of Example #5

aaaaaa
1111

The compressed state of Example #5 is 10 characters from the original non-compression total of 100 characters in length.

6. Biology

Biology is the study of nature and as such is a part of the systematic atomistic axiomization of processes found within living things. These natural grammars, or laws, has mathematical corollates that parallel process found in the physical and engineering disciplines. The use of a compression algorithm of a sequential string is a natural development of such a process as can be seen in the compression of both DNA and RNA genetic codes.

6.1 DNA

DNA or Deoxyribonucleic acid, is a linear polymer made up of specific repeating segments of phosphodiester bonds and is a carrier of genetic information [10]. There are four bases in DNA; adenine, thymine, guanine and cytosine [11].

The use of a compression algorithm for sequences of DNA.

Definitions:
A = Adenine
T = Thymine
G = Guanine
C = Cytosine

Example #A

ATATGCGCATATCGCGTATATATATATA

The compression algorithm will use a specific focus on TA and GC DNA sequences in Example #A.

Key Code

TA = 6 characters

GC = 2 characters

Compress Example #A

ATAT<u>GC</u>ATATCGCG<u>TA</u>

5

The compressed DNA sequence is 16 characters from the original non compression total of a 28 character length.

6.2 RNA

RNA, or Ribonucleic acid, translates the genetic information found in DNA into proteins [12]. There are four bases that attached to each ribos [13].

Definitions:
A = Adenine
C = Cytosine
G = Guanine
U = Uracil

Example #B

AUAUCGCGAUAUCGCGUAUAUAUAUAUAGCGC

The compression algorithm will focus on specific RNA sequences.

Key Code

UA = 6 characters

GC = 2 characters

Compress Example #B

AUAUCGCGAUAUCGCGUAGC

The compressed RNA sequence is 20 characters in length from the original non-compression total character length of 32.

7. Nanotechnology

The development and discovery of nanometer scale structures, ranging from 1 to 100 nanometers, to transform matter , energy and information on a molecular level of technology [14].

7.1 Synthetic Biology
Within the field of synthetic biology is the development of synthetic genomics that uses aspects of genetic modification on pre-existing life forms to produce a product or desired behavior in the life form created [15].

The following is a DNA sequence of real and 'made up' synthetic sequences.

Definitions:
A = Adenine
T = Thymine
G = Guanine
C = Cytosine
W= *Watson
K = *Crick

*Note: Made up synthetic DNA.

Example #C

TATAGCGCWKWKATATCGCGKWKWKWKWKWKW

Key Code

AT = 2 characters

CG = 2 characters

KW = 6 characters

Compressed Example #C

TATAGCGCWKWKATCGKW

The compressed synthetic DNA sequence is 18 characters from the original non-compression character total of 32.

Summary

The paper has addressed the use of a compression algorithm for use in various radix based number systems in the fields of chemistry, biology and nanotechnology. The compression algorithm in both the universal and specific format have successfully reduced long and short sequences of strings to very compressed states and function well in both random and non-random sequential strings.

References

1. Shannon, C.E., *Bell Labs. Tech. Jour.* 27, 379-423 and 623-656 (1948).

2. Tice, B.S., "The analysis of binary, ternary and quaternary based systems for communications theory", Poster for the SPIE Symposium on Optical Engineering and Application Conference, San Diego, California, August 10-14, 2008.

3. Kotz, S. and Johnson, N.I., *Encyclopedia of Statistical Sciences,* John Wiley & Sons, New York (1982).

4. Martin-Lof, P., "The definition of random sequences", *Information and Control*, 9, pp. 602-619 (1966).

5. Tice, abide.

6. Richards, R.K., *Algorithmic Operations in Digital Computers*, D. Van Nostrand Company, Princeton, NJ, (1955).

7. Moore, J.A. *McGraw-Hill Encyclopedia of Chemistry*, McGraw-Hill Publishers, New York (1993).

8. Wikipedia. "Polymer". Wikipedia, September 4, 2010, p. 1.
   Website: http://en.wikipedia.org/wiki/Polymers.

9. Wikipedia. "Copolymer". Wikipedia, September 4, 2010, pp. 1-5.
   Website: http://en.wikipedia.org/wiki/Copolymers.

10. Lutter, L.C., "Deoxyribonucleic acid". In *McGraw-Hill Encyclopedia of science & technology*.
    McGraw-Hill Publishers, New York. Pp. 373-379 (2007).

11. Lutter, abide., p.374.

12. Beyer, A.L. and Gray, M.W., "Ribosomes". In *McGraw-Hill Encyclopedia of science & technology*.
    McGraw-Hill Publishers, New York. Pp. 542-546 (2007).

13. Beyer, abide., p. 542.

14. Drexler, K.E., "Nanotechnology". In *McGraw-Hill Encyclopedia of science & technology*.
    McGraw-Hill Publishers, New York, pp. 604-607 (2007).

15. Wikipedia. "Synthetic genomics". Wikipedia, September 4, 2010, p. 1.
    Website: http://en.wikipeida/wiki/Synthetic-genomics.